DONKEY FARMING FOR

BEGINNERS

A Complete Guide To Raising, Breeding, And Caring For Healthy Rustics For Profit And Sustainability

Holden bodhi

Contents

DISCLAIMER

The information provided in this book, is intended for educational and informational purposes only. The content is based on research, personal experiences, and general knowledge about farming. It is not intended to substitute professional advice or expert consultation. Readers are encouraged to seek professional guidance when implementing any practices or techniques discussed in this book.

The author and publisher make no representations or warranties of any kind regarding the accuracy, applicability, or completeness of the contents of this book. Any reliance you place on such information is strictly at your own risk. The author and publisher shall not be held liable for any damages, losses, or injuries resulting from the use of the information provided.

Additionally, the author does not endorse, recommend, or affiliate with any individual, product, service, website, organization, or brand mentioned or referenced in this book. Any such references are solely for informational purposes, and no warranty or guarantee is implied. The inclusion of these references does not imply any endorsement or partnership by the author.

By reading this book, you acknowledge and accept that the author and publisher are not responsible for any consequences arising from your use of the information provided

CHAPTER ONE

Overview Of Donkey Farming

Animal lovers and small-scale farmers are finding that donkey farming is an interesting and lucrative agricultural endeavor. For thousands of years, people from many cultures have relied on these docile, resilient creatures for various tasks, such as companionship, field preparation, and transportation. Donkey farming provides a simple and accessible introduction to livestock farming for novices, as it does not require the same level of effort or complexity as raising horses or cattle. Donkeys are a great option for novices since they require little upkeep, are adaptable, and are simple to care for.

Donkey farming has been more involved in recent years than it was in the past. It is now involved in the production of donkey milk, which is in demand due to its health benefits, as well as eco-tourism and therapy animals. The principles of donkey care, breed selection, and the many advantages of having these animals should be understood whether you plan to raise donkeys for personal use, as working animals, or as a component of a larger farming enterprise.

This tutorial will cover the fundamentals of donkey farming, from choosing the best breed to realizing the various ways that donkeys can make farming more rewarding and sustainable.

An Overview Of Cattle Raising

Despite being more specialized than other livestock businesses, donkey farming provides a variety of opportunities for small-scale and hobby farmers alike. Donkeys are adaptable animals that can be raised for milk production, labor, and breeding. They are renowned for their kind nature, robust work ethic, and ability to persevere in trying circumstances. They are especially enticing to novices who might not have much expertise in cattle production because of these qualities.

Compared to larger animals like horses or cows, donkeys require less room and resources, thus they may be kept on smaller pieces of land. They are simple to feed in most climates because they can graze on a wide range of grasses and fodder. In addition, compared to other livestock, donkeys are comparatively disease-resistant and have fewer health issues, which can drastically lower veterinary expenses and the need for frequent intervention.

To ensure that the donkeys in the farming practice flourish, proper care, nutrition, and shelter must be provided. Given their gregarious nature, donkeys should ideally be housed in couples or small groups. Providing kids with mental stimulation and enrichment, like safe toys or varied locations, is beneficial for their well-being because they also have a natural curiosity and intelligence.

Farmers should understand the varied applications of donkeys in diverse farming environments in addition to providing basic care. There are many ways to turn donkey farming into a lucrative and pleasurable business, whether your goals are to utilize them for transportation, labor in agricultural fields, or investigating niche industries like donkey milk manufacturing.

Advantages Of Bringing Up Donkeys

Raising donkeys has several advantages, especially for newcomers or those wishing to add some variety to their farm animals. Compared to other animals, donkeys are easy to handle because of their kind and quiet disposition. They can be wonderful farm animal companions or even family pets because of their kindness and patience.

The hardiness of donkeys is one of the main benefits of rearing them. They can survive in both desert and temperate climates because of their great degree of climate adaptation. Compared to other farm animals, they are less prone to disease and injury because of their robust immune systems, resilient hooves, and innate ability to forage for food. This results in fewer maintenance and veterinary expenses.

Donkeys are also incredibly adaptable creatures. Light to moderate labor, like pulling wagons, lifting cargo, and even plowing small fields, can be done with them. For small-scale farmers or those without easy access to machinery, this makes them indispensable. Donkey manure may also be a fantastic source of fertilizer, assisting farmers in maintaining healthy, fertile soil for agricultural development.

The market for donkey milk, which is prized for its nutritional richness and hypoallergenic qualities, is expanding, which is another new advantage of donkey farming. Donkey milk provides farmers with an additional source of revenue because it is utilized in skincare products and as a food supplement. Donkeys are also gaining popularity in the field of animal-assisted therapy, where their kind disposition benefits therapeutic initiatives for adults and children with emotional difficulties or disabilities.

Raising donkeys ultimately pays off financially and practically. These animals offer a special opportunity for people interested in lucrative and ecological farming, whether they are utilized for labor, friendship, or the production of valued goods like donkey milk.

Breeds Of Donkeys That Are Good For Novices

For novices in particular, selecting the appropriate breed is essential to success in donkey farming. There are various donkey breeds, and each has its own special traits, stature, and disposition. While certain breeds are better suited for hard work, others are better for producing milk or providing companionship. Novices must choose a breed that is amiable, adaptable, and minimal maintenance.

The miniature donkey breed is among the most well-liked varieties for novices. These little, easy-to-manage donkeys, who are originally from the Mediterranean, are renowned for being calm and amiable. They are very well-liked as companion animals and are frequently seen in therapy or petting zoos. Because they require less room and feed, miniature donkeys are perfect for small farms or hobby farms.

The Standard Donkey, which is larger than the miniature variety but still manageable for individuals who are new to farming, is another excellent option for novices. These donkeys are frequently employed for menial tasks like pushing carts or carrying weights. Their easygoing attitude and ability to adapt to many situations make them an ideal choice for farmers seeking low-maintenance, flexible livestock.

Poitou Donkeys are a great breed to think about if you're interested in producing milk. Poitou donkeys are native to France and are distinguished by their long, shaggy coats and kind disposition. Compared to other donkey breeds, they are larger, which is perfect for milk production. However since they need more room and feed, they might not be the greatest choice for farmers with tight budgets.

The goal of your donkey farming business, the resources you have at your disposal, and the amount of time you have to devote to animal care should all be taken into account when choosing a breed. An easy-to-manage and well-cared-for breed will be the starting point for a prosperous and pleasurable donkey farming endeavor.

CHAPTER TWO

Putting Your Donkey Farm In Order

A donkey farm is an interesting project that needs to be carefully thought out and planned. Although donkeys are hardy creatures, their production and general well-being are heavily influenced by the conditions in which they live. This part will walk you through how to choose the best site, set up your farm properly, make sure your donkeys have enough shelter and fence, and get their surroundings ready.

Choosing A Site And Layout For The Farm

One of the most important choices you will make for your donkey farm is where to locate it. The perfect location should be roomy, safe, and easily accessible. Donkeys are prone to hoof issues if they stand in damp circumstances for an extended period of time, so look for a place with sufficient drainage to avoid waterlogging.

Climate-Related Issues

Your chosen location's climate has a big impact on donkey health. Although they can tolerate a variety of weather situations, donkeys flourish in temperate climes. If you live somewhere really hot, make sure there are places under the shade where the donkeys can get some shade. On the other hand, if you live in a colder area, think about installing windbreaks to shield them from strong winds.

Space Needs

Donkeys need enough room to move about and graze. Generally speaking, you should provide each donkey at least one acre. For the sake of their mental and physical health, this area permits them to partake in natural activities like grazing and socializing. Establish distinct spaces for shelter, grazing, and feeding to guarantee a well-planned farm layout.

Availability

Make sure that you, any guests, and veterinarians can all readily reach your farm. This includes any trails or well-kept roads that lead to the farm. To increase management efficiency, take into account the distance to essential services like veterinarian care and feed suppliers.

Needs For Shelter And Fencing

Options for Fencing

If you want to keep your donkeys contained and safe, fencing is essential. A strong fence that is five feet high and constructed of wire or wood is the ideal kind of fencing for donkeys. Due to their natural curiosity, donkeys occasionally manage to break free from their enclosure when they spot something intriguing outside of it. It is not advisable to use barbed wire because it could hurt your donkeys.

Gate Observations

Make certain that every gate is both safe and simple to open. For ease of use by trucks and equipment, gates should be at least four feet wide. For larger spaces, it's also desirable to use double gates to allow for movement without stressing out the animals.

Requirements for Shelter

Providing your donkeys with enough shelter is vital to their well-being. A straightforward barn or three-sided shelter can provide weather protection. Make sure there are no drafts and adequate ventilation in the shelter. Since donkeys like to lie down on dry land, floors should be sturdy and hygienic.

Comfort and Bedding

For comfort and moisture absorption, line the shelter with wood shavings or straws. Clean the shelter on a regular basis to keep it hygienic and disease-free. Additionally, this will keep the donkeys happy and healthy.

Setting Up The Scene For Donkeys

It takes more than simply buildings to provide your donkeys with a comfortable living space. You also need to take into account their social and psychological needs.

Pasture Administration

Since donkeys are grazers by nature, a healthy pasture is essential to their diet. Make sure there is plenty of grass and no harmful weeds in the pasture. To avoid overgrazing and to give the grass time to recuperate, rotate the grazing areas.

Water Source

It is vital to have access to fresh, clean water. Install easily refilled and cleaned water troughs. To make sure water sources are uncontaminated, check them frequently.

Requirements for Socialisation

Being gregarious creatures, donkeys live best in groups. To meet the social needs of donkeys, it is ideal to maintain a minimum of two of them together. This will lessen the likelihood of loneliness and related behavioral issues.

Activities for Enrichment

Give your donkeys toys or other something to play with to keep their minds engaged. They can be entertained and encouraged to play with simple objects like balls or cones, which lowers the likelihood of behaviors connected to boredom.

In summary

A donkey farm requires careful design and consideration of several aspects, such as layout and location, fencing, and environmental preparedness. You may create the conditions for a fruitful and satisfying experience with donkey farming by creating a secure, cozy, and engaging environment. We will go into more detail on donkey management and care in the upcoming parts to make sure they prosper on your farm.

CHAPTER THREE

Selecting Appropriate Donkeys

Choosing the appropriate donkeys for your farm is one of the most important decisions you'll make when you start donkey farming. Donkeys are unusual creatures with distinctive qualities that may or may not make them suitable for labor, breeding, or friendship. This section will cover the important things to think about when choosing donkeys, where to get healthy examples, and how to recognize the telltale symptoms of a healthy donkey.

Things To Take Into Account When Choosing Donkeys

Why We Farm

The first step in selecting the appropriate donkeys for your agricultural endeavor is to ascertain its goal. Donkeys can be used as pack animals, companions, protectors of other animals, or even for breeding. Different types or breeds of donkeys may be needed for each application. For instance, if you plan to utilize donkeys for labor-intensive tasks like moving materials or carrying weights, you should think about selecting strong and resilient breeds like the American Mammoth Jackstock. On the other hand, smaller breeds like the Miniature Mediterranean Donkey can be more appropriate if you're looking for a friend or pet.

Features of the Breed

Making an informed choice requires having a thorough understanding of the breed's traits. The temperaments, sizes, and maintenance requirements of various donkey breeds vary. While smaller breeds are frequently easier to care for and may require less room, larger breeds may be more suited for employment. Investigate different breeds to learn about their advantages and disadvantages. The Standard Donkey, for example, is a great option for novices due to its amiable nature and versatility.

Size and Age

The donkey's size and age are important considerations. Although older donkeys may arrive with established behaviors and training, they may also have health concerns. Younger donkeys may require more socialization and training. Larger breeds might require more room to wander and graze, so keep that in mind as well. Limited breeds, on the other hand, can flourish in limited meadows.

Socialisation and Temperament

Donkeys are gregarious creatures that do well in packs. Think about the disposition of the donkeys and how well they will fit into your farming setting when choosing them. Seek out companionable and manageable donkeys. If at all feasible, watch the donkeys in their natural habitat to evaluate how they behave and interact with people and other animals. By keeping an eye on this, you can make sure the donkeys you choose are both healthy and well-behaved.

Where To Get Well-Being Donkeys

reputable breeders

Reputable breeders are among the greatest places to buy healthy donkeys. Seek out breeders who are well-known in the donkey farming industry and who specialize in the breed you are interested in. A reputable breeder has to be open to your inquiries, give you details about the health history of the donkeys, and let you tour their facility. Make sure you are dealing with a reliable breeder by reading internet reviews and requesting referrals.

Shelters and Rescue Organisations

Adopting donkeys from animal shelters or rescue groups is a great alternative as well. Adopting a donkey can be a fulfilling experience since many donkeys in shelters need loving homes. Donkey rescue organizations frequently evaluate the health and behavior of the animals, giving you important details about the requirements and history of the donkey. Before choosing, be sure the organization is well-respected and that the animals are receiving proper care.

Auctions of Livestock

Although they carry some risk, livestock auctions might be a good way to discover donkeys. Even while you might be able to obtain healthy donkeys for less money, you should still perform in-depth examinations and ideally bring along an experienced person to help you assess the animals. To make an informed decision, be ready to enquire about the donkey's past, present, and behavior.

Examining Indications Of A Fit Donkey

Outward look

Examining a donkey's outside look is crucial when choosing one. A donkey in good health should have a glossy coat devoid of rashes, parasites, or bald areas. Look for a well-formed body and bright, clear eyes. The donkey should be balanced, with even hooves and a straight back. Avoid a donkey that seems drowsy, has a dull coat, or exhibits symptoms of malnourishment.

Actions

Another important determinant of the donkey's health is its behavior. A healthy donkey ought to be gregarious, inquisitive, and vigilant. Watch how it behaves with humans and other animals. Your experience farming could be complicated if the donkey exhibits fear, aggression, or withdrawal. These behaviors could be caused by underlying health or behavioral difficulties.

Veterinary Documentation

Veterinary documents should be requested before completing your purchase. Documentation of the donkey's immunizations, deworming, and medical treatments should be provided by a reliable breeder or rescue group. You can learn more about the donkey's medical history and prevent future health issues by looking through these data.

Examinations of Health

If at all feasible, have the donkey examined by a veterinarian prior to taking it home. A veterinarian can conduct a comprehensive checkup to rule out frequent illnesses and determine whether the donkey is suitable for farming. All donkeys need to have routine health examinations, but this is especially important when adding additional animals to your farm.

In summary

A crucial first step in any agricultural endeavor is selecting the appropriate donkeys. This involves giving careful thought to several aspects, such as the intended use, breed traits, and health indicators. A good donkey farming experience can be built around a careful evaluation and investigation of prospective donkeys. Keep in mind that donkeys require long-term care, and choosing healthy, suitable animals will make your farming experience more fulfilling and fruitful.

CHAPTER FOUR

Nutrition And Feeding

Nutrition and feeding are essential aspects of donkey care that have a big impact on the health, output, and general well-being of the animals. Novice donkey farmers need to comprehend their food requirements to maintain robust and healthy animals.

Knowing Your Dietary Needs, Donkey Style

As herbivores, donkeys need a diet that is balanced and suited to their particular physiological requirements. Donkeys, in contrast to horses, have adapted to live well on low-nutrient feed and are adept at using fibrous materials. Roughage processing is a function of their evolved digestive systems and is essential to their general well-being.

Providing high-fiber, low-protein fodder for donkeys is one of the main things to take into account while feeding them. The majority of their diet needs to consist of grass hay, including timothy or orchard grass. Alfalfa hay is frequently excessively rich for donkeys, and if fed in excess, it can cause obesity or other health problems.

Donkeys also require less energy than horses do, therefore it's important not to overfeed them with concentrates or grains. A small amount of grain may occasionally help them, but it should only be administered under certain circumstances, including during nursing or the healing process from an illness.

It's also critical to understand that donkeys are susceptible to several health problems, such as obesity and laminitis, or inflammation of the foot. It is therefore essential to keep an eye on their body condition score (BCS) to make sure they continue to maintain a healthy weight. BCS scores of 4 to 6 indicate an optimal weight range, and the scale runs from 1 to 9. You can modify their feeding schedule as needed with the assistance of routine evaluations of their health.

Feed And Supplement Types

There are a variety of feeds and supplements available to help ensure that donkeys are fed a balanced diet.

1. Hay: As previously indicated, the main food source for a donkey should be hay. Alfalfa should be fed in moderation and grass hay is the recommended option. To prevent respiratory problems, make sure the hay is clear of mold, dust, and other impurities.

2. Pelleted Feed: To guarantee a balanced diet, pelleted feeds designed especially for donkeys are available and can be an excellent choice. Seek for fiber-rich, low-calorie pellets that are low in sugar and starch.

3. Grain: Donkeys can eat grains, but only in small amounts should be provided. When choosing grains, steer clear of sweet feeds and go for high-fiber choices. If your donkey needs more energy—during pregnancy, for example— speak with a veterinarian to figure out how much is right.

4. Minerals and Vitamins: Supplementing with minerals and vitamins is crucial, particularly for donkeys whose main diet is hay. By preventing vitamin shortages, these supplements promote general health. Because donkeys' demands are different from horses', look for goods made especially for these animals.

5. Salt Licks: To stay hydrated and preserve good health, donkeys need access to salt. Providing loose salt or mineral salt blocks will guarantee kids get enough sodium.

6. Fresh Produce: Carrots, apples, and other fruits are popular treats for donkeys. These ought to be consumed in moderation, though, as an excess of sugar can result in obesity and other health problems.

Schedules For Watering And Feeding

Donkeys' health and well-being depend on having a regular feeding and watering regimen. It is impossible to negotiate for access to fresh, pure water. A donkey's daily water intake might range from five to ten gallons, based on its size, degree of activity, and surroundings. To avoid contamination, make sure the water troughs are cleaned regularly.

Feeding Schedule

Providing hay and/or food twice a day is a common feeding routine. This is a potential strategy:

1. Morning: Give them a daily quantity of hay (around 1.5% of their body weight) along with any supplements that

they may require. They will have the energy they need to start the day thanks to this.

2. Afternoon: Verify the water supply and top it off if necessary. If you decide to give grain or pellets, give them in tiny amounts in the afternoon, particularly if they are nursing or pregnant.

3. Evening: Give the hay a second helping. Given that donkeys are grazers by nature and should have access to food all day long, this is crucial for them.

Keeping an eye on intake

It's critical to keep an eye on how much water and nourishment your donkeys are getting. Weigh their feed and hay regularly to make sure they are getting the proper amounts. Watch how they eat; a sudden shift could point to a health problem that needs to be attended to by a veterinarian right away.

Considering the Seasons

Seasonal variations may require adjusting feeding schedules. Donkeys may need more water in the sweltering summer months and more forage in the winter to stay warm. Be ready to modify feed kinds and amounts in response to variations in environmental factors or activity levels.

CHAPTER FIVE

Veterinary Care And Health In Donkey Farming

Maintaining a healthy and successful farm while raising donkeys depends on taking good care of them. Despite being resilient creatures, donkeys are susceptible to various illnesses and conditions that may lower their quality of life. You can make sure your donkeys live long and healthy lives by being aware of common health issues, setting up a schedule for checkups and immunizations, and consulting a veterinarian frequently. Each of these crucial topics is covered in detail in this section.

Typical Donkey Illnesses And How To Avoid Them

Like other livestock, donkeys can contract a variety of illnesses, many of which can be avoided with good management and care. Reducing health risks on your farm requires knowledge of these prevalent diseases and skills in their prevention.

1. Hoof Issues

Hoof problems are one of the most prevalent health problems in donkeys. Conditions that can cause pain and hinder mobility, including laminitis, abscesses, and enlarged feet, are common in donkeys. Laminitis is an inflammation of the delicate laminae in the hoof, which can be brought on by an unhealthy diet, being overweight, or working too much on hard surfaces. Regular foot clipping (every 6 to 10 weeks), eating a balanced diet, and making sure donkeys have access to soft ground for hoof resting are examples of preventive practices.

2. Colic

Donkeys may also be at risk for the painful digestive disorder known as colonic. It can be brought on by abrupt dietary adjustments, subpar nutrition, or dehydration. An enlarged belly, lack of appetite, restlessness, and rolling about on the ground are some of the symptoms. Donkeys should always have access to clean water, and a regular, high-fiber diet, and should not have abrupt changes in their feeding schedule to prevent colic.

3. infections of the respiratory system

Donkey respiratory conditions, like pneumonia or equine influenza, can have a serious negative effect on health. Poor air quality in barns and stables can further spread these infections, as can close contact with other affected animals. Coughing, nasal discharge, and difficulty breathing are among the symptoms. To prevent respiratory infections, it is important to keep living areas clean and well-ventilated, isolate sick animals, and administer the proper immunizations.

4. Insect parasites

Donkeys are frequently afflicted with both internal and external parasites, including ticks, mites, and worms. For instance, worms can cause digestive problems, poor coat condition, and weight loss. The parasite burden in your donkeys can be managed by routine deworming, clean pastures, and grazing area rotation. Ticks and other external parasites can be controlled by giving the donkeys regular brushings and administering insecticidal medications as required.

Routine Vaccinations And Health Examinations

One of the most important aspects of donkey husbandry is establishing a plan for routine immunizations and health checks. By identifying early symptoms of disease, these examinations facilitate prompt treatment and help avert more serious health problems.

1. Body Condition Points

A quick and easy way to keep an eye on your donkey's health is to find out its body condition score (BCS). This entails touching important regions including the shoulders, spine, and ribs to cover muscle and fat. A BCS that is too low can indicate underweight and a BCS that is too high puts the donkey at risk of obesity, which can cause laminitis and other health issues. You can modify your donkey's diet and training regimen by monitoring their body condition score (BCS).

2. Dental Well-being

To avoid enlarged teeth, which can cause eating difficulties, weight loss, and infections, donkeys need to have regular dental treatment. It is advised to examine the donkey's teeth at least once a year for jagged edges or uneven wear. Throughout their lifetimes, donkeys' teeth grow continuously. If their teeth aren't appropriately worn down by grazing, they may need to be floated, which is a procedure when a veterinarian files the teeth to make their surface smoother.

3. Schedule of Vaccinations

One essential element of preventative healthcare is vaccination. The three main vaccinations for donkeys are usually rabies, equine influenza, and tetanus. You might need to have extra vaccinations against diseases like equine herpesvirus or West Nile Virus, depending on where you live and the dangers in the area. A suggested immunisation regimen depending on the unique circumstances of your farm can be given by your veterinarian.

4. Consistent Control of Parasites

Choosing the right deworming schedule for your donkeys depends on regularly testing their feces for internal parasites. Developing a strategic plan for parasite control with the assistance of a veterinarian is crucial, as overuse of dewormers may result in resistance. Lice and mites are examples of external parasites that can be managed with consistent grooming, topical medicines, and clean bedding.

Getting Ongoing Care From A Vet

It is vital to establish a solid rapport with a veterinarian who is knowledgeable about donkey health. A skilled veterinarian can offer advice on regular maintenance, handling emergency cases, and any specialized care that the donkey might need over its lifetime.

1. Regular Examinations

Making routine veterinary appointments for your donkeys helps to guarantee that any new health issues are detected early. During these visits, the donkey's medical history, dental health, and records of its deworming and vaccinations can all be reviewed. To maintain optimum health, a veterinarian can also provide nutritional recommendations, particularly for elderly donkeys or those with unique dietary requirements.

2. Being Ready for Emergency Care

Apart from providing regular treatment, your veterinarian is a valuable asset in times of need. Having a veterinarian on standby can be lifesaving when dealing with an unexpected colic episode or a devastating injury.

While you wait for veterinary care, it can be useful to carry a first-aid kit filled with necessities like bandages, antiseptic solutions, and painkillers for small injuries that may occur.

3. Plans for Health Management

It's crucial to work with your veterinarian to develop customized health management programs for each donkey, especially those with severe health issues. This plan may include dietary and exercise guidelines as well as ongoing monitoring for long-term ailments like metabolic disorders or arthritis. You can make sure that your donkeys get the greatest care possible for the duration of their lives by adhering to a customized plan.

In summary, keeping your donkeys healthy requires tight coordination with a veterinarian, preventative measures, and routine health examinations. Being proactive and aware of your donkeys' requirements will help you lower the chance of illness and maintain your animals in top shape for many years to come.

CHAPTER SIX

Raising Donkeys

Having healthy, productive donkeys for breeding may be a fulfilling endeavor, providing the chance to grow useful animals for companionship, transportation, and agriculture. Careful preparation and a deep understanding of donkey genetics, medicine, and management techniques are essential for successful breeding. Three main topics are covered in this section: choosing breeding partners, tending to the pregnancy and foaling, and caring for the newborn donkeys.

Choosing Breeding Pairs

Choosing the appropriate breeding couples is the first and most important stage in donkey breeding. The health, disposition, and physical attributes of the progeny will be greatly influenced by the genetic features of the dam (a female donkey) and the father (a male donkey). A few things to think about when selecting breeding couples are as follows:

1. Vitality and Health: Only sound donkeys ought to be used for breeding. A comprehensive veterinary examination is necessary to guarantee that neither of the animals has any hereditary illnesses, ailments, or disorders that might be passed on to their progeny. Before mating, donkeys with common health concerns such as respiratory illnesses, dental disorders, and hoof problems should be ruled out.

2. Temperament: Although donkeys are often thought of as being calm and patient, each one has a different temperament. Choose paired parents who exhibit steady, dependable conduct to enhance the probability of generating children with well-mannered dispositions. This is especially crucial if the donkeys are going to be used in human-human interactions, either as therapy or pack animals.

3. Size and Conformation: When choosing breeding partners, size and conformation—the physical makeup of the donkey—are crucial factors to take into account. Breeding donkeys with extreme size variations might cause issues with pregnancy and delivery. Breeding couples should ideally be similar in size and conformation to provide a more seamless pregnancy and foaling experience. The donkey's ability to carry weights and pull carts is also impacted by conformation.

4. Breed considerations: Donkeys come in a variety of breeds, each with their special traits. For instance, the Miniature Mediterranean donkey is valued for its smaller stature and mild disposition, whereas the American Mammoth Jackstock is a huge breed frequently utilized to produce powerful working animals. Choose couples that support any particular objectives you may have for the children, such as breeding donkeys for pack work, riding, or friendship.

5. Age: Three to ten years old is the best range for donkeys to breed. While elderly donkeys may have decreased fertility and higher foaling risks, younger donkeys may not

be completely matured, which could cause health issues during pregnancy.

A good donkey breeding program starts with carefully choosing breeding pairings, which increases the chance of having offspring that are physically sound, well-tempered, and healthy.

Care During Pregnancy And Foaling

It is crucial to provide Jenny, the female donkey, the attention she needs to ensure a healthy pregnancy and successful foaling after breeding has taken place and it has been determined that she is pregnant. Due to their lengthy gestation period—which can last anywhere from 11 to 14 months—donkeys require constant observation for the health and welfare of the baby.

1. Nutrition: Jenny's nutritional requirements rise throughout pregnancy, particularly in the later stages. Give her a diet heavy in quality fodder, such as hay and pasture, and well-balanced with vitamins and minerals. Steer clear of overfeeding as this may cause problems during foaling. Continually supply fresh, clean water to ensure she stays hydrated.

2. Veterinary Care: To keep an eye on the health of the pregnant rabbit and make sure the fetus is growing normally, routine veterinary examinations are crucial. Prenatal care also includes key topics such as vaccinations and parasite control. To determine a safe vaccination and deworming schedule for pregnant donkeys, speak with a veterinarian.

3. Exercise: Donkeys who are pregnant can benefit from moderate exercise. Give the buck enough room to roam around in a secure, contained space; this will keep her muscles toned and lower the possibility of issues during foaling. Steer clear of demanding activities and make sure she has plenty of room to graze and walk.

4. Foaling Preparation: It's critical to get a hygienic and peaceful birthing space ready as the foaling date draws near. This could be a safe pasture with an adequate cover or a specialized foaling stall. Ensure there are no risks present and that there is an ample supply of clean bedding. Donkeys frequently want to give birth in peace, and they may foal at night, so keep an eye on things and keep the area well-lit.

5. Signs of Impending Foaling: The jenny will exhibit physical symptoms such as an expanded udder, a softening of the pelvic muscles, and restlessness as the time of foaling approaches. Keep an eye out for these symptoms and be ready to help if needed, although most donkeys fall easily. Call a veterinarian right away if Jenny appears distressed or if the foaling procedure is taking too long.

Handling Young Donkeys

The foal's survival and health depend heavily on what happens in the initial hours and days after birth. Taking good care of a newborn donkey involves tending to its needs as soon as it is born, keeping an eye on its health, and making sure the foal and the jenny bond and grow together.

1. Care After Birth: The foal should be naturally cleaned by the jenny by licking it, which promotes breathing and

circulation. Within the first hour, make sure the foal is upright and breathing normally. You might need to help by gently wiping the foal with fresh, dry towels if the jenny isn't cleaning it.

2. Nursing and Colostrum Intake: The foal needs to start nursing as soon as possible after birth to get its hands on colostrum, the first milk the jenny produces which is packed with vital nutrients and antibodies. The foal receives passive immunity from the colostrum, shielding it from illness in its early years. You might need to help the foal nurse or get advice from a veterinarian if it is having trouble.

3. Health Monitoring: Throughout the first few days, pay special attention to the foal's behavior and development. Active and inquisitive, healthy foals should be routinely nursed. Inspect for indications of weakness or illness, such as weakness, diarrhea, or trouble standing, and seek veterinary attention if necessary.

4. Bonding with Jenny: The emotional and physical growth of the foal depends on the relationship that exists between Jenny and her offspring. Give them plenty of time to get to know one another and try not to take them apart too soon in the beginning. The foal will be naturally protected and cared for by Jenny, but she will make sure they are in a secure space where they may interact freely.

5. Vaccinations & Hoof Care: Discuss a suitable vaccination and deworming program with your veterinarian as the foal grows. To guarantee healthy growth, you should

also keep an eye out for any anomalies in the foal's hooves and arrange for routine hoof treatment.

You can create the conditions for healthy, well-behaved animals that will flourish in the future by taking good care of your newborn donkeys. Success in donkey breeding requires a thorough approach to breeding, pregnancy care, and foal management.

CHAPTER SEVEN

Training And Managing Donkeys

To guarantee that your donkeys are obedient, secure, and productive, you must train them. Donkeys are distinguished from horses and other animals by their intelligence, patience, and cautious disposition, which calls for a special method of handling and training. Establishing a strong base in handling strategies, training approaches, and trust-building can benefit both you and your donkeys.

Fundamental Training Methods For Donkeys

The first step in training a donkey is to become familiar with its temperament and natural behavior. Compared to other farm animals, donkeys are more autonomous thinkers and can come out as obstinate or uncooperative. But their cautious mentality frequently results in their so-called intransigence. Donkeys can be trained to carry out duties, obey orders, and engage with handlers in a safe manner if the proper methods are used.

Positive Reinforcement: Among the best methods of training for donkeys is this approach. Donkeys are more likely to repeat desirable behaviors when they receive treats or affection when they perform appropriately, as opposed to punishment-based strategies that may instill fear and resistance. Giving a treat to a donkey that follows you on a lead, for instance, when it walks without straying, encourages it to continue doing so.

Clicker Training: Another useful technique for training donkeys is clicker training. With this method, a tiny clicker is used to emit a sound each time the donkey carries out a desired activity. After hearing this sound, you get rewarded with a goodie. The click will eventually be associated by the donkey with a favorable result, supporting the reinforcement of learned behaviors. Donkeys can be trained to respond to specific commands, such as elevating their hooves for foot care or remaining motionless during grooming, with good results when using clicker training.

Beginning with Basic Commands: It's important to begin with the fundamentals when training a donkey. Early in the training phase, commands like "come," "stop," and "stand" ought to be introduced. To help your donkey learn and retain orders, repetition is essential. To prevent overtaxing the animal, begin training sessions briefly and progressively increase the duration and level of difficulty of each session.

Training for Desensitisation: Like many other animals, donkeys can become sensitive to novel situations, sounds, and sensations. The goal of desensitization training is to gradually expose individuals to different stimuli. This could involve encountering unfamiliar creatures, hearing startling noises, or traveling across various surfaces. Your donkey will become more at ease and situation-adaptive if you introduce these things to it securely and peacefully.

Managing Donkeys Safely

As with any animal handling, safety comes first, and donkeys are no exception. Although donkeys are normally friendly animals, mishandling can result in mishaps or make the animal afraid or reluctant. It's essential to learn proper handling methods for the donkey's welfare as well as the handler's safety.

How to Approach Your Donkey: Make sure the donkey can see you and approach it slowly and calmly at all times. Because donkeys have strong survival instincts, they may perceive any abrupt movement or loud noise as a threat. It's important to approach them directly or from the side so they can see you well. Stealthily approach them from behind to avoid surprising them and provoking defensive behaviors like kicking.

Leading Your Donkey: Being able to lead a donkey is a fundamental handling ability that enables you to transport them from one location to another. Make sure the halter you use fits the donkey's head comfortably and snugly. Always stand to the left of the donkey while leading, and grasp the lead rope firmly but loosely. Never put the rope around your hand since the donkey can suddenly pull away and hurt you. Walk with assurance but with gentleness, letting the donkey follow at its speed.

Lifting and Examining Hooves: Donkeys require proper foot care, and being able to raise and examine their hooves is a necessary component of safe handling. Be certain that a donkey is at ease and comfortable before attempting to lift its hoof. Apply light pressure as you move your hand down the donkey's leg, causing it to shift its weight and raise its hoof on its own. Avoid stressing or hurting the donkey by being patient and without forcing it to obey. Routine maintenance, like cleaning and trimming, will be considerably easier with regular hoof handling and lifting.

Using Safe Restraints: A donkey may need to be restrained to perform some operations, such as medical treatment or foot clipping. It's crucial to use compassionate and safe restraints in these situations to avoid hurting anyone. By tying your donkey with a lead rope in a safe, fast-releasing knot, you can prevent it from straying and make sure it can be quickly released in an emergency. Tie the rope loosely enough so the donkey won't feel imprisoned and start to panic. A holding pen or donkey chute may be utilized for more intricate treatments to safely immobilize the animal.

Developing Confidence With Your Donkeys

Establishing trust is essential for effective training and handling of donkeys. Donkeys are naturally wary creatures, and the degree to which they trust their handler determines how cooperative they will be. Building trust takes time, patience, and consistency and is a progressive process.

Time Spent Together: Simply spending time with your donkeys is the first step in developing trust. Frequent, easygoing interactions—such as sitting in the pasture or giving the donkey a hand-feed—that include no demands of the animal to let it get to know you better. These times of companionship greatly contribute to the development of trust because donkeys are gregarious animals who love to form bonds with their owners.

Gentle Handling: Calm, gentle handling is well-received by donkeys. To prevent frightening your donkey, move slowly and steadily when interacting with them. Regularly giving your donkey affectionate touches or brushing aids in strengthening a favorable link with your presence. Observe the donkey's body language; if it appears tense or anxious, give it some room to settle before moving on.

Patience and Consistency: Repetition and consistent interaction are the best ways for donkeys to learn. Be patient; hurrying the process can result in disappointments, especially in the early training sessions. To help your donkeys know what to expect, feed, handle, and train them according to a regular schedule. With time, your donkeys' comfort level and level of trust will increase, which will facilitate training and handling.

Reward-Based Interactions: As was already mentioned, fostering trust can be achieved through providing positive reinforcement. Give your donkey a treat or some verbal praise each time they behaves nicely under control or engage in pleasant interactions. This promotes cooperative behavior and strengthens your relationship with the donkey. Reward or punishment should be avoided as this undermines trust and makes training harder in the future.

In conclusion, a combination of appropriate methods, safety precautions, and trust-building are necessary for successful donkey handling and training. On your farm, you may raise cooperative, well-mannered donkeys that are a pleasure to work with by getting to know your donkey's character, utilizing positive reinforcement, and maintaining a safe and stable environment.

CHAPTER EIGHT

Needs For Housing And Shelter In Donkey Farming

Although donkeys are resilient creatures, it is crucial to provide them with sufficient housing and shelter to maintain their productivity, well-being, and health. A well-thought-out shelter guarantees comfort shields them from inclement weather, and lets you effectively maintain their hygiene. The essentials of building a donkey shelter, controlling bedding and hygiene, and incorporating climate factors into your donkey care regimen will all be covered in this part.

How To Create A Donkey Shelter

Space Needs

The first and most crucial factor to take into account while building a donkey shelter is room. Given their sociable nature, donkeys require plenty of room to roam around, rest, and socialize with other donkeys. The number of donkeys you intend to house will determine the size of the shelter, but generally speaking, each donkey needs about 150 square feet of covered space. The shelter should also be tall enough (about ten feet) to provide adequate airflow and room for mobility.

Position and Direction

The success of your donkey shelter greatly depends on where it is placed. It should ideally be situated on higher land to avoid flooding when the rainy season arrives. The interior of the shelter will stay dry and toasty if it is orientated with the opening facing away from the direction of the prevailing winds. For the ease of the donkeys and to reduce their stress levels, it is also crucial to make sure that grazing pastures and water supplies are easily accessible.

Material Decisions

Safe and long-lasting materials should be used to construct the shelter. Materials that are frequently used are concrete, steel, and wood. The walls must be sturdy enough to shield the donkeys from the elements and provide adequate airflow to keep the inside dry and mold-free. Strong, waterproof roofing materials that can survive inclement weather are what are needed. Because they are long-lasting, metal roofs are frequently chosen; however, ensure that they are insulated to keep the interior from becoming too hot on hot days.

Light and Ventilation

Enough ventilation is necessary to keep donkeys from developing respiratory problems. The shelter should keep drafts out, but it should also let enough air circulate, preventing the accumulation of moisture and urine-derived ammonia. Natural ventilation can be achieved through large windows, vents, or apertures in the top walls. Donkeys also benefit from sunlight, so building the shelter with this feature in mind can lift their spirits and enhance their general well-being.

Bedding And Organising Tidiness

Selecting the Ideal Mattress

Another essential component of your donkey's housing is the bedding. In addition to offering comfort and insulation, proper bedding also keeps the shelter dry and clean. Common bedding materials include sawdust, wood shavings, and straw; of these, straw is the most popular because it's readily available and reasonably priced. Since straw bedding can easily get wet and dirty and cause respiratory illnesses and hoof rot, it should be changed regularly.

Daily Schedules for Cleaning

For the donkeys' health, the shelter must be kept clean. Donkeys are generally clean creatures and like dry, orderly environments. To keep dangerous bacteria and parasites from growing, the shelter should be cleaned every day by removing damp bedding and manure. Every day, fresh bedding needs to be added to the donkeys' sleeping quarters to guarantee their comfort. To keep the area hygienic, manure should be composted or disposed of in an environmentally responsible way.

Timetable for Deep Cleaning

Every few weeks, in addition to routine upkeep, the shelter should be thoroughly cleaned. This entails taking out all of the bedding, giving the walls and flooring a thorough cleaning, and waiting for the shelter to completely dry before putting in fresh bedding. Diseases can be stopped from spreading by using disinfectants that are safe for animals. Preserving the shelter's cleanliness and dryness also lessens the possibility of drawing flies, who can irritate and sick donkeys.

Management of Wastes

Keeping your donkeys healthy is mostly dependent on proper waste management. Wet bedding and accumulated manure can draw bugs and encourage the spread of illness. To lessen the likelihood of rodent infestations, regularly remove garbage from the shelter and maintain the area surrounding it free. An environmentally friendly way to supply your farm with organic fertilizer is by composting manure.

Climate-Related Issues In Donkey Care
Tolerance to Heat and Cold

Although they can adapt to a variety of settings, donkeys are sensitive to extremes of heat or cold. To avoid heat stress in warm climates, shelters should include plenty of shade and ventilation. As dehydration, fatigue, and heatstroke can result from overheating, make sure they have access to fresh water and don't spend too much time in the sun. Donkeys should be able to hide in the shady sections of the shelter during the warmest parts of the day.

Insulation becomes essential in colder areas. Although a donkey's natural coat offers some protection from the cold, extra warmth and shelter are required in the winter. To avoid moisture, the shelter needs to be well-ventilated even without any drafts. To assist in keeping the interior at a suitable temperature, you can add insulation or windbreaks. To keep the donkeys warm during the winter, provide them with more bedding.

Defence Against Snow and Rain

Donkeys may have discomfort and health problems from rain and snow, especially if they stay wet for long periods. To facilitate the easy drainage of snow and precipitation, the shelter's roof should be slanted. Furthermore, make sure the shelter is waterproof because donkeys require dry conditions to prevent diseases in their hooves and rain rot. The shelter should be sturdy enough to support the weight of the snow in locations where there is a lot of it, and the donkeys should have unobstructed passageways.

Air Quality and Humidity

The ventilation of the shelter requires extra care in areas with high humidity. Elevated humidity levels can generate moist surroundings that encourage the development of bacteria and fungi, resulting in respiratory issues and skin illnesses in donkeys. Maintaining airflow and lowering humidity levels inside the shelter can be facilitated by installing suitable ventilation equipment, such as windows or ceiling fans.

Getting Used to Seasonal Shifts

It's critical to modify the shelter in accordance with seasonal variations. To keep donkeys cool throughout the summer, you might need to install fans, offer more shade, or even utilize misting systems. Windbreaks, heat lamps, or even heated water troughs can help keep them comfortable in the winter. Your donkeys will stay healthy all year long if you monitor the weather forecast and make any necessary alterations to the shelter.

Donkey farming is essentially about giving your donkeys a good home and shelter. You may establish a secure and healthful space that supports their well-being by taking into account their requirements for comfort, space, and climate. To guarantee that your shelter is a comfortable place for your donkeys throughout the year, regular upkeep and cleaning are equally crucial.

CHAPTER NINE

The Socialisation And Behaviour Of Donkeys

Recognizing The Behaviour Of Donkeys

Donkeys are intelligent, gregarious animals with a wide variety of behaviors driven by their innate social structures and instincts. It is essential to comprehend these behaviors in order to manage donkeys effectively and maintain a positive rapport between workers and their charges.

1. Herd dynamics and instincts

Since donkeys are herd animals by nature, social groups are essential to their survival. They develop close relationships with their friends in the wild and depend on one another for support and company. Isolated donkeys may experience stress, anxiety, or depressive symptoms. When establishing a donkey farm, it is important to recognize this innate tendency. To provide donkeys the social interaction they require to flourish, it is usually advised to keep at least two of them together.

2. Signals of Communication

Donkeys use a range of vocalizations and body language to convey messages. Their body language can convey feelings of hostility, obedience, or contentment, and their braying can convey anything from joy to distress. A donkey, for instance, that stands with its tail swishing and its ears pulled back is probably agitated or threatened. A donkey, on the other hand, indicates comfort and curiosity if it comes towards you with relaxed ears and a soft nickel. It will be easier for you to efficiently manage and care for your donkeys if you pay attention to these indications.

3. Playful Actions

Foals, or young donkeys, playfully engage in activities that are vital to their socialization. They can develop physical confidence and social hierarchies through play, which includes kicking, chasing, and even play-fighting. Playing with toys like balls or ropes or just associating with one another, adult donkeys also love to play. Playtime helps maintain donkeys' mental stimulation and lessen boredom-related behavioral problems.

Introducing Humans And Other Animals To Donkey Socialisation

A vital component of donkey farming is socialization. Donkeys who have been socialized appropriately are easier to manage and are more likely to form wholesome bonds with people and other animals.

1. Gradual Exposure to Humans

It's crucial to approach donkeys cautiously to socialize them with people. Spend some time with them first, observing without pressuring them to talk. Let them watch you so they can become used to your voice and smell. This method lessens anxiety and fosters trust. You can reinforce positive interactions by giving food or introducing mild caressing after the donkey appears comfortable.

2. Regular Management and Instruction

The secret to socializing donkeys is to handle them consistently. Make feeding, grooming, and exercise routines. Include fundamental training methods that can help you establish your leadership role, including leading on a halter or obeying orders. Treats and other forms of positive reinforcement, such as praise, can encourage donkeys to interact with you and pick up new skills. Frequent communication will help you and your donkey develop a closer relationship.

3. Presenting Additional Animals

Use caution when acclimating donkeys to other animals. Commence with a gradual introduction that lets both animals watch each other from a distance. Encourage closer contact after they seem at ease, making sure that everyone feels secure. Due to their territorial nature and potential for defensive behavior in response to threats, keep a watchful eye on their interactions. Gradual introductions aid in avoiding unpleasant experiences that can impair socialization in the future.

Taking Care Of Behavioural Issues

Donkeys are like other animals in that they can have behavioral issues. To successfully address these behaviors, it is imperative to identify their underlying causes.

1. Identifying Stress Signals

Stress can be brought on by several things for donkeys, including environmental changes, a lack of company, or disease. Acknowledging the indicators of stress, like rapid breathing, altered eating patterns, or excessive vocalization can enable you to intervene to ease their suffering. Reducing stress-related behaviors requires having a stable environment, regular social interactions, and access to quality healthcare.

2. Putting Training Methods into Practice

Training is commonly used to address behavioral problems including fear, stubbornness, or violence. Positive reinforcement strategies can be used to reward positive behavior and discourage negative behavior. For example, teaching a donkey to wait quietly for food if they become hostile during feeding times can be beneficial. Establishing firm boundaries and unambiguous instructions will help you effectively control their behavior.

3. Getting Expert Assistance

If behavioral issues don't go away with your efforts, you could want to hire a professional equestrian trainer or animal behaviorist. These professionals can offer specially designed plans for dealing with particular problems and enhancing the general well-being of your donkeys. Regular veterinarian examinations can also rule out any underlying medical conditions that can be linked to behavioral disorders.

CHAPTER TEN

Tasks And Roles For Working Donkeys

For generations, donkeys have played a crucial role in both agriculture and transportation. They are useful assets on farms due to their strength and adaptability, which they use for a variety of jobs that increase productivity. This section will examine the various functions that donkeys can perform on farms, as well as their usage as pack animals and other special contributions.

Utilizing Donkeys In Agriculture

Donkeys are remarkable animals for agricultural labor. They can carry out tasks that would require a lot of work for people thanks to their power and endurance.

Setting Up the Land

Donkeys help prepare the land, which is one of their main jobs on a farm. They can be used to till the soil, plow fields, and get ground ready for planting. Utilizing a donkey's inherent strength allows farmers to minimize the amount of manual labor needed. This is especially helpful for smaller farming enterprises when purchasing machinery might not be financially feasible.

Carriage of Products

Donkeys are useful for moving products throughout the farm in addition to preparing the ground. They are perfect for transporting equipment, harvested crops, and other goods from one place to another because of their strong construction, which enables them to lift huge loads. This ability to transport people is especially helpful in steep or uneven terrain where cars may have difficulty.

Grasping and Managing Insects

Additionally, donkeys can aid with natural insect control. They can aid in weed control without the use of herbicides by being allowed to graze on fields. This lowers farming expenses related to weed control while simultaneously encouraging a healthier ecosystem.

Ecological Farming Methods

Farming with donkeys encourages sustainability. Their presence helps to move agriculture towards a more environmentally friendly model by reducing reliance on fossil fuels. Furthermore, farmers can utilize more of their property because donkeys can flourish on food that may not be suited for other livestock.

As Pack Animals, Donkeys

It has long been known that donkeys make dependable pack animals. They are a good fit for this function because of their distinct temperament and physical attributes.

Power and Sturdiness

Donkeys' strength is one of the main benefits of employing them as pack animals. They are capable of transporting large weights over long distances—up to 20% of their body weight. Because of their stamina, they can travel for extended periods without the need for regular rests, which makes them perfect for carrying cargo over difficult terrain.

Comfort and Load Distribution

It's critical to appropriately distribute the weight while packing donkeys. This avoids harm while also guaranteeing the donkey's comfort. Their capacity to carry goods efficiently can be improved with the use of cushioned equipment and proper saddle practices. To make sure their donkeys are comfortable carrying loads, farmers should make an effort to train them in this skill.

Flexibility in Pack Work

Donkeys can be utilized for a variety of pack tasks, including hauling goods and tools or delivering produce to markets. Because of their versatility, they can be used on a variety of farms, regardless of size.

Examining Additional Functions For Donkeys

Donkeys can take on a variety of roles that improve a farm's overall efficiency and sustainability in addition to regular farm labor and pack chores.

Friendliness and Safety

Donkeys can be great animals to have around other cattle. They are a great asset to farms that have sheep, goats, or cattle because of their ability to defend against predators. The mere presence of donkeys in the pasture can serve as a deterrent to possible threats.

Animals used for therapy

Donkeys are good for therapy work because of their gentle nature. They can offer people consolation and emotional support in a variety of environments, such as hospitals, schools, and rehabilitation facilities. Their kind nature and serene demeanor can support therapeutic programs and enhance mental health.

Roles in Education

Donkeys have gained popularity in educational settings in recent years. Children can learn about agriculture, responsibility, and animal care from them. Donkeys represent animal welfare and sustainable farming methods in this capacity.

Soil Health and Biodiversity

Donkeys can enhance biodiversity on a farm by living in harmony with other creatures and vegetation. Their grazing practices can support the upkeep of thriving pasture ecosystems. Their dung also acts as a natural fertilizer, improving the soil and encouraging the growth of healthy crops.

CHAPTER ELEVEN

Essentials Of Donkey Farming Business

Overview Of Donkey Farming

The fascinating business of donkey farming has grown in popularity due to its adaptability and sustainability. Donkeys are a great option for both small and large farms since they are resilient animals that can live in a variety of settings. It's important to comprehend the fundamentals of starting and maintaining a profitable donkey farm before venturing into the realm of donkey farming. This section covers important topics such as the kinds of donkeys, facilities that are required, and basic care needs.

Different kinds of donkeys

Donkeys come in a variety of breeds, and each has special qualities that suit them for a particular use. The Standard Donkey, Miniature Donkey, and Mammoth Donkey are the most popular varieties.

The Standard Donkey is a multipurpose animal that is utilized for both friendship and labor. Mammoth donkeys are frequently employed in agricultural activities because of their enormous size and power, whereas miniature donkeys are popular as pets and in therapy because of their gentle nature. You can choose the best breed for your agricultural objectives by being aware of the traits and qualities of the various breeds.

Crucial Infrastructure

Appropriate facilities are necessary for a successful donkey farm to guarantee the animals' health and well-being. For them to be able to move around and graze, donkeys require a large, safe enclosure. A perfect shelter would offer them ventilation and protection from inclement weather. A dry, spotless space is essential for feeding and sleeping. To stop the spread of illness, take into account the requirement for separate enclosures for breeding, and weaning foals, and sick animals.

Daily Maintenance Needs

Even though donkeys are considered low-maintenance pets, they nevertheless need daily attention. Feeding, grooming, and wellness checks are examples of basic care. Hay, oats, and freshwater make up a well-balanced diet. Vaccinations and routine hoof trimming are essential for their well-being. Developing a care regimen for your donkeys will not only keep them healthy but also strengthen the relationship between you and the animals, which is essential for successful farming.

Starting A Donkey Farm: What It Costs
First Invested

A donkey farm's initial expenses can vary according to the location, size, and objectives of the enterprise. Typically, the initial outlay covers the cost of equipment, fencing, shelters, and land. Conducting market research is crucial because land costs might differ greatly depending on one's geographic location. Donkey safety and confinement depend on the type and length of fencing needed, which can be expensive.

Acquiring Donkeys

Donkey purchases can vary significantly in price based on the breed and age of the animal. Standard and Mammoth donkeys can cost anywhere from $500 to $3,000. Miniature donkeys can cost anywhere from $1,500 to $5,000. When creating your budget, don't forget to account for continuing expenses related to breeding, medical care, and feeding. You may choose your herd more wisely if you are aware of the usual market prices.

Function-related Costs

Running a donkey farm involves significant operating costs in addition to the original investment. Feed, vet care, grooming supplies, and facility upkeep are examples of regular expenses. Feed can take up a large amount of your cash, so finding premium grains and hay at reasonable pricing is crucial. The general health of the donkeys depends on veterinary care, which includes both routine examinations and emergency treatment.

Reserve Money

It's a good idea to set up a contingency reserve for unforeseen costs. Unexpected medical expenses, facility repairs, or market swings may all be covered by this fund. As a general guideline, allocate at least 10% to 15% of your overall budget for unforeseen expenses to make sure you can handle unexpected difficulties without endangering the farm's profitability.

Earning Applications For Donkeys

Breeding

Breeding donkeys is one of the most profitable parts of the business. You may produce superior babies that appeal to a variety of markets by breeding donkeys for sale if you have the necessary information and resources. To guarantee that the breeding process produces healthy, desired offspring, an understanding of genetics and health factors is essential.

Laboring Animals

Due to their strength and endurance, donkeys have been utilized as labor animals for millennia in a variety of agricultural jobs. They can be employed for load carrying, field plow work, and even animal herding. Especially in remote locations where traditional farming practices are still used, you can make a good living by renting out your donkeys for employment.

Travel and Counselling

The employment of donkeys in rehabilitation and tourism is growing. In rural tourism, donkey rides can be a well-liked attraction since they provide guests with an unusual experience. Furthermore, because of their gentle disposition and capacity for human connection, donkey therapy programs are becoming more and more popular. You may reach a larger audience and diversify your revenue sources by providing these services.

Trail riding and packing

Utilizing donkeys for packaging in rural or mountainous places is another lucrative option. They are perfect for outdoor trips because they can carry stuff and are great hiking companions.

You can increase your business opportunities by bringing in visitors and outdoor enthusiasts by planning guided trail rides or packing expeditions.

Promoting And Selling Services Or Donkeys

Finding the Right Target Markets

Your donkey farming business must have successful marketing if you want to succeed. Decide which markets are your target first.

Think about factors related to age, interests, and geography. Families seeking pets, farmers in need of work animals, and companies interested in therapeutic programs are examples of potential clients. Your chances of success will increase if you target these groups with your marketing efforts.

Establishing a Virtual Identity

It is essential to have an online presence in the modern digital world. Make a polished website that highlights your farm, the donkeys you raise, and the services you provide. Share updates, pictures, and anecdotes about your donkeys on social media. Creating interesting material can draw in followers and future clients, fostering a sense of community around your farm.

Creating Connections with Local Communities

Making connections with local communities might open up a lot of beneficial doors for you to sell your services. Participate in agricultural exhibitions, farmer's markets, and local fairs to advertise your services and display your donkeys. Developing connections with nearby companies can result in joint ventures and recommendations. The success and visibility of your farm can be greatly impacted by building a strong local presence.

Set Your Services and Donkey Prices

When selling services or donkeys, it is important to have competitive prices. To find out the going rates for donkeys and associated services in your area, perform some market research. You can draw in more clients by setting up several price tiers according to the age, breed, or level of training of your donkeys. Prospective customers will respect you more if you are open and honest about your pricing and communicate the worth of your services.

Customer Support and Input

Finally, delivering top-notch customer service is crucial for sustained success. Create a feedback system to get information from clients regarding their experiences. Reacting to criticism, whether constructive or positive, shows that you are dedicated to excellence and development. Customers who are happy with your service are more likely to recommend you to others and come back for more, which can help your donkey farming business expand.

CONCLUSION

For those who are new to agriculture and want to explore something different, donkey farming offers a gratifying and exciting opportunity. Raising donkeys is a rewarding experience that has several financial benefits in addition to personal benefits. These amazing animals are well-known for their resilience, intelligence, and kind disposition, which makes them the perfect choice for a variety of farming tasks, such as transportation, land maintenance, and animal care.

The fundamentals of donkey farming have been covered in this tutorial, including choosing the best breed, setting up acceptable living quarters, comprehending dietary demands, and identifying medical issues. Beginners may give their donkeys the greatest care possible according to these recommendations, which will ultimately result in a more fruitful and satisfying farming experience.

Furthermore, interacting with regional networks and communities for donkey farming can provide priceless resources and assistance. Those who are passionate about these amazing animals can develop a sense of camaraderie by exchanging experiences with other farmers, which can improve knowledge and abilities. Success in farming, as in all endeavors, is largely dependent on ongoing education, flexibility, and a strong sense of responsibility for the well-being of the animals under your care.

Keep these things in mind as you start your donkey farming adventure: patience and observation. Recognizing the distinctive personalities of your donkeys will help you and your pets form closer bonds. You may create a successful donkey farming business that enhances your life and the community with happiness, productivity, and sustainability if you put in the necessary time, effort, and love.

Learning to Farm Donkeys

Knowing the Fundamentals of Donkey Farming

An increasingly common business that draws people and families interested in agriculture and animal husbandry is donkey farming. Understanding the foundations of donkey care and management is essential for novices. Being hardy animals, donkeys can survive in a variety of settings, including suburban backyards and rural farms. However, successful farming requires an understanding of their unique demands.

Choosing the Correct Breed

Choosing the appropriate breed is the first step in establishing a donkey farm. There are various donkey breeds, and each has special traits and uses. For example, the Standard and Mammoth kinds of donkeys are frequently utilized for work or as pack animals, whereas the Miniature Mediterranean donkey is ideal for friendship and compact areas. Selecting the breed that best suits your farming objectives can be made easier if you do your research on the many varieties.

Environment and Housing

Donkeys need to live in comfortable surroundings, thus this must be done. They need a spotless, dry area with sufficient protection from inclement weather. They'll stay happy and healthy in a basic barn or stable with enough airflow. Donkeys also require access to pasture or grazing grounds in order to exercise and eat. Since donkeys are considered to be highly intelligent, fencing needs to be strong and tight to prevent escapes.

dietary and feeding

Donkey production and health depend heavily on proper nutrition. Grass pasture grass, high-fiber hay, and specially prepared donkey feed are usually components of a balanced diet. There should always be access to fresh water. Donkeys are prone to obesity and associated health problems, so it's crucial to prevent overfeeding them. Knowing what donkeys need to eat can make them healthier and more productive.

Medical Treatment and Upkeep

Regular medical attention is essential to donkey farming. Developing a rapport with a veterinarian knowledgeable about donkey care can have a big impact. Hoof trimming, dental care, and routine immunizations should all be included in your maintenance regimen. Additionally, you can keep your donkeys healthy and happy by keeping an eye out for any indications of disease or suffering.

Creating a Community

Finally, making connections with other donkey farmers or agricultural communities might help you navigate your new endeavor by offering information and assistance. Online discussion boards, regional get-togethers, and workshops can provide insightful information, useful advice, and a sense of community among other donkey aficionados.

www.ingramcontent.com/pod-product-compliance
Lightning Source LLC
Chambersburg PA
CBHW052339220526
45472CB00001B/491